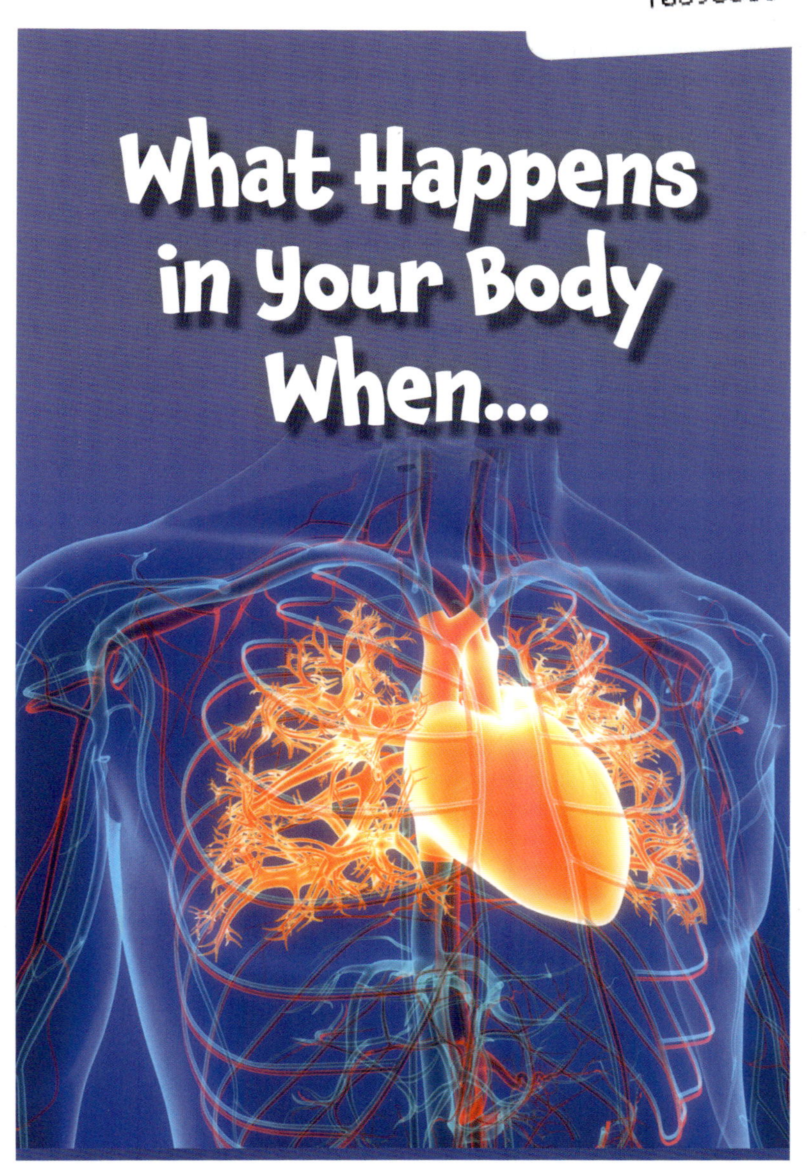

Amanda Jackson Green

Consultants

Darrin Lunde
Collection Manager
National Museum of Natural History

Cheryl Lane, M.Ed.
Seventh Grade Science Teacher
Chino Valley Unified School District

Michelle Wertman, M.S.Ed.
Literacy Specialist
New York City Public Schools

Publishing Credits

Rachelle Cracchiolo, M.S.Ed., *Publisher*
Emily R. Smith, M.A.Ed., *SVP of Content Development*
Véronique Bos, *VP of Creative*
Dani Neiley, *Editor*
Robin Erickson, *Senior Art Director*
Jill Malcolm, *Senior Graphic Designer*

Smithsonian Enterprises

Avery Naughton, *Licensing Coordinator*
Paige Towler, *Editorial Lead*
Jill Corcoran, *Senior Director, Licensed Publishing*
Brigid Ferraro, *Vice President of New Business and Licensing*
Carol LeBlanc, *President*

Image Credits: p.14 AMI IMAGES/Science Source; p.15 Alamy Stock Photo; p.16 Juan Gaertner/Science Source; p.21 Getty Images; all other images from iStock and/or Shutterstock.

Library of Congress Cataloging in Publication Control Number: 2024024230

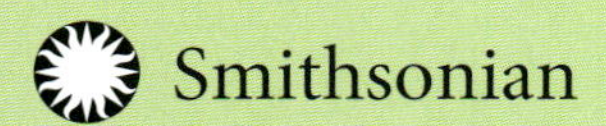

5482 Argosy Avenue
Huntington Beach, CA 92649
www.tcmpub.com
ISBN 979-8-7659-6857-4

Printed by: 51497
Printed in : China

Table of Contents

The Amazing Human Body

Think of a group of people you know. Maybe you're picturing your family, your sports team, or the other students in your class. No matter which group you imagine, the people in it may have a lot of physical differences. Height, skin tone, and features (such as hair color or eye color) likely vary among each person in the group. Although their bodies look different on the outside, they have a lot in common on the inside. Human bodies share almost all their experiences and functions. The differences are tiny compared to what they have in common.

Every minute of every day, the tissues, organs, and systems that make up our bodies perform lots of tasks. These jobs keep us alive and help us interact with the world around us. A person's body keeps their heart beating, blood pumping, and lungs breathing. The person does not have to think about doing those things—their body does them automatically. This is all thanks to the basic parts and complex systems in their body.

When we understand how the human body works, we can better care for ourselves and others. We can learn how to stay healthy and heal from illness or injury. This knowledge can support us as we grow older. Learning about the body's many complex parts and systems can be a lifelong journey!

All About Healthy Cells

At the smallest level, the basic building block of the human body is the **cell**. All living **organisms**—including tiny **bacteria**, plants, animals, and humans—are made up of cells. Examining cells and how they work helps us understand all the amazing things human bodies can do.

Let's start with the parts of a cell. Most cells in the human body have three main parts. First, the membrane is the outermost layer. It acts like a gatekeeper. The membrane allows some substances, such as oxygen, to enter a cell. This layer can also block substances, such as acids, from getting in. Next, the nucleus is at the center of a cell. The nucleus is like the mastermind of a cell, continually giving instructions. The nucleus can tell a cell when to grow, multiply, or release waste. Finally, like the glue that holds everything together, **cytoplasm** fills the spaces between all the parts of a cell. Together, these three parts perform key jobs for cells.

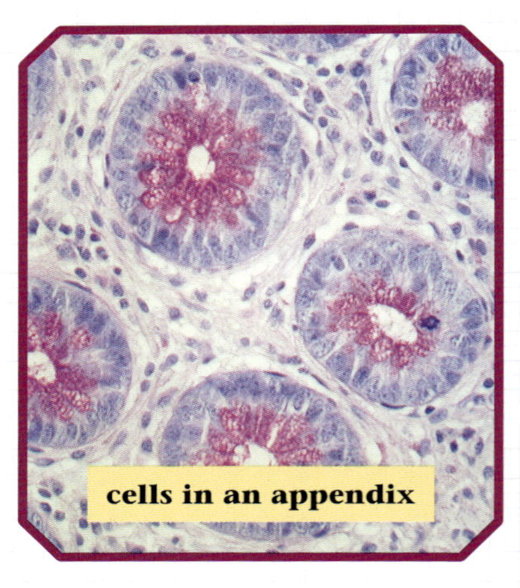
cells in an appendix

Cells also contain several types of **organelles**. Organelles perform specific functions for a cell. For example, oval-shaped mitochondria take in oxygen and nutrients, transforming them into energy. Ribosomes are another type of organelle. They look like small dots, and they produce proteins. Proteins are vital in every cell. They give a cell structure and help it complete important tasks. Finally, globe-shaped lysosomes act like a cell's maintenance crew. They remove waste to help a cell maintain optimal performance.

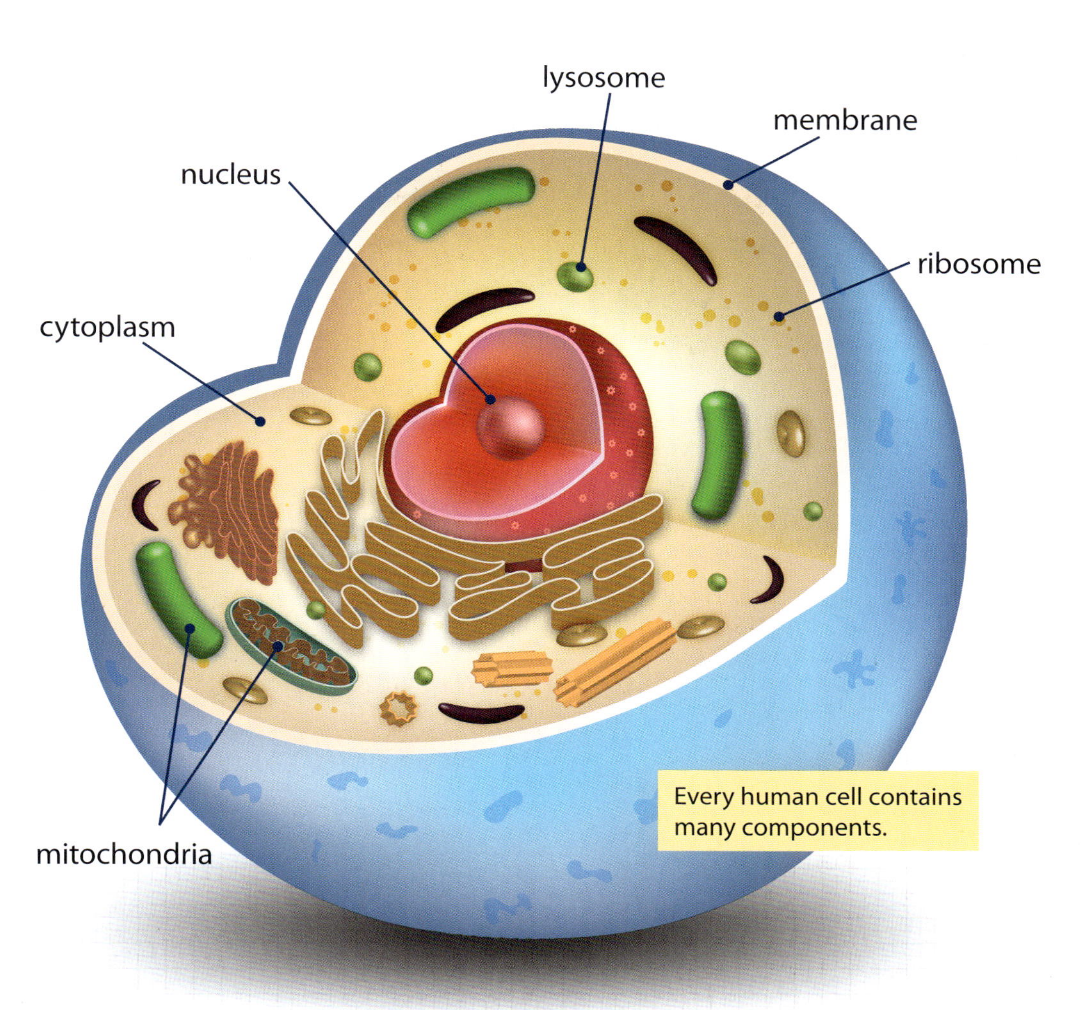

Every human cell contains many components.

Chromosomes can only be seen under a microscope.

FUN FACT

Inside the nucleus of a cell, you'll find chromosomes. These are structures that carry DNA, or deoxyribonucleic acid. DNA contains all the information about how living things look and function. DNA is a tiny molecule made of two linked strands. Its twisted, ladderlike shape is called a *double helix*.

DNA

More than 200 types of cells are found in human bodies. Human cells have varying sizes and **densities**. Most cells in humans are so small that they can only be seen under a microscope. Cells are in a continuous state of change and **replication**. Different types of cells divide and multiply at different speeds. These factors make it difficult for scientists to know the number of cells in one person. However, recent estimates put the number at a whopping 36 trillion cells for an adult male.

All this change and replication can be a good thing when it comes to healing. In normal functioning, cells are constantly renewing the human body toward health and wellness.

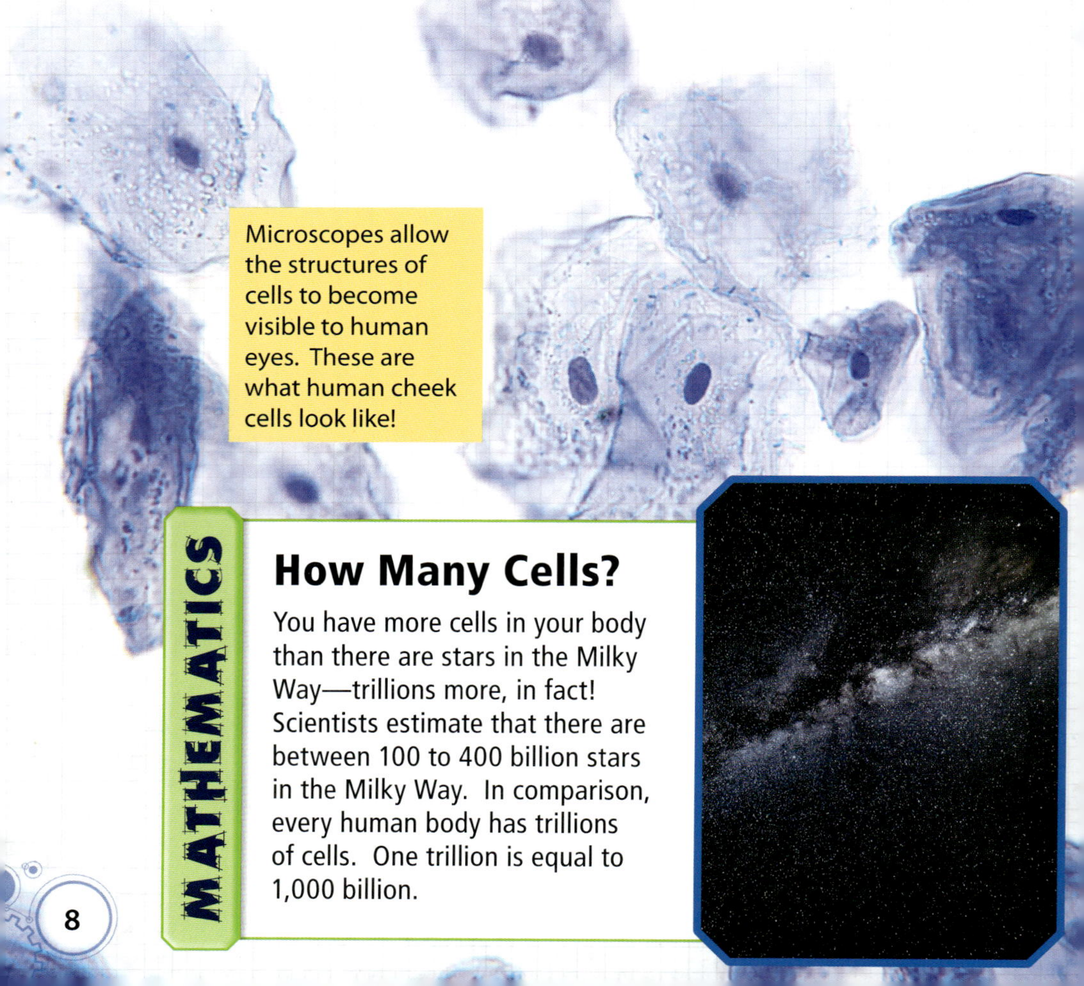

Microscopes allow the structures of cells to become visible to human eyes. These are what human cheek cells look like!

MATHEMATICS

How Many Cells?

You have more cells in your body than there are stars in the Milky Way—trillions more, in fact! Scientists estimate that there are between 100 to 400 billion stars in the Milky Way. In comparison, every human body has trillions of cells. One trillion is equal to 1,000 billion.

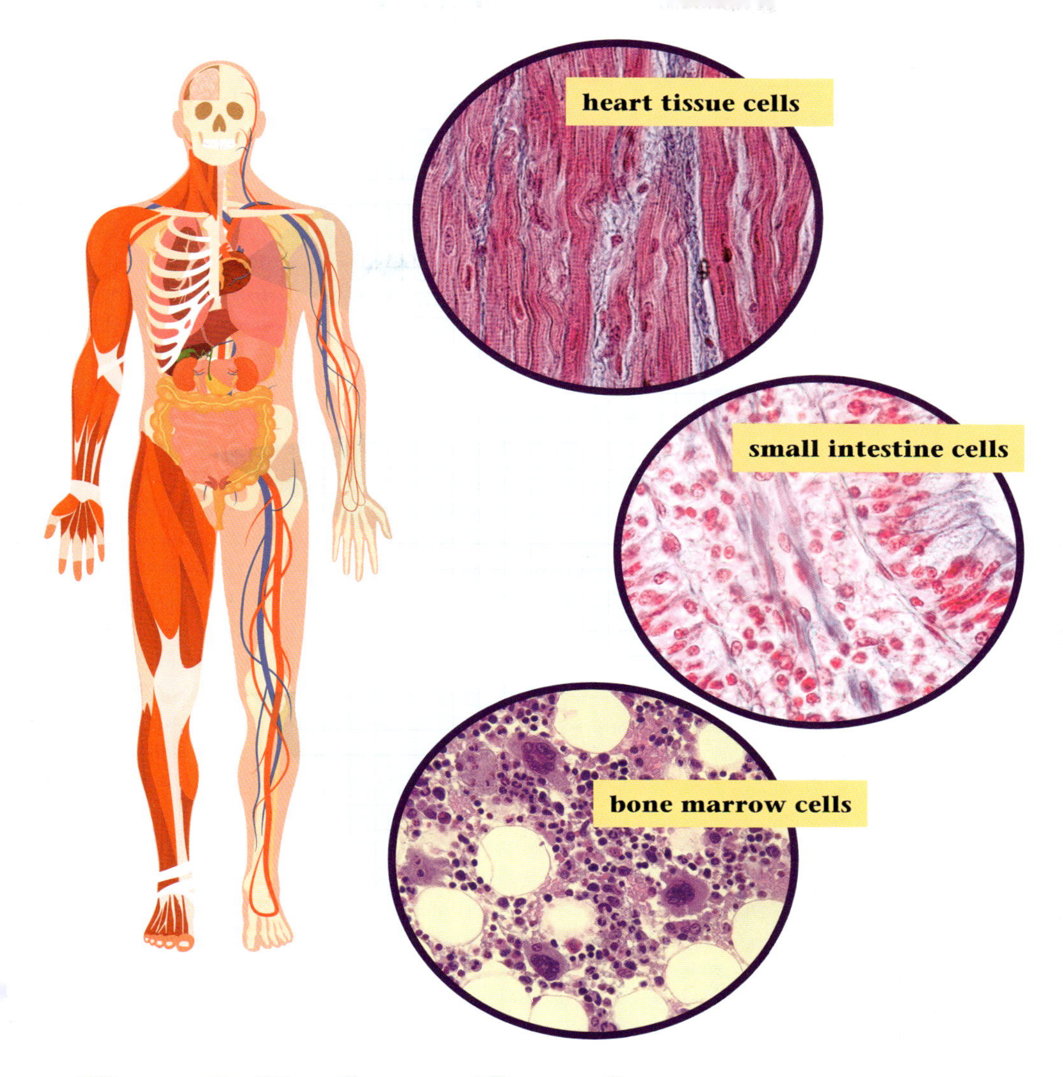

How Cells Come Together

In human bodies, various combinations of cells join to form tissues. Some examples of tissues are blood, **muscles**, fat, and bones. Organs, such as the liver and heart, are made up of several kinds of tissues. Each organ performs a specific function. For example, the lungs take in oxygen and release carbon dioxide. Together, organs and tissues make up bodily systems. These systems complete more complex tasks, such as digesting food or controlling bodily movements. Systems keep human bodies stable and healthy, all while supporting the tasks of daily life. Understanding these systems allows us to learn what happens in the body in a healthy state.

A Deeper Look at Systems

A healthy human body is like a fine-tuned car. It has different parts that work well in harmony. The systems in a person's body combine forces to make life's activities go smoothly. Here are some of the key systems and how they work in the human body.

Structural systems provide support for a body. Made up of 206 bones, the skeletal system gives a body its shape. Bones also act as guards, protecting the body's organs from harm. For example, the ribcage protects the heart and lungs. A close friend to the skeletal system is the muscular system. This system is made up of hundreds of muscles. Many of these muscles attach directly to bones. Tissues in muscles stretch and shorten to produce movements.

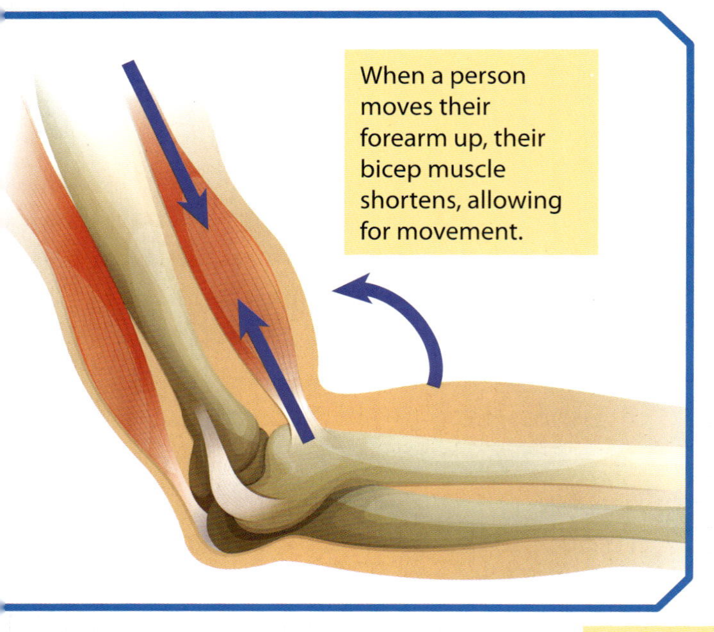
When a person moves their forearm up, their bicep muscle shortens, allowing for movement.

muscular and skeletal systems

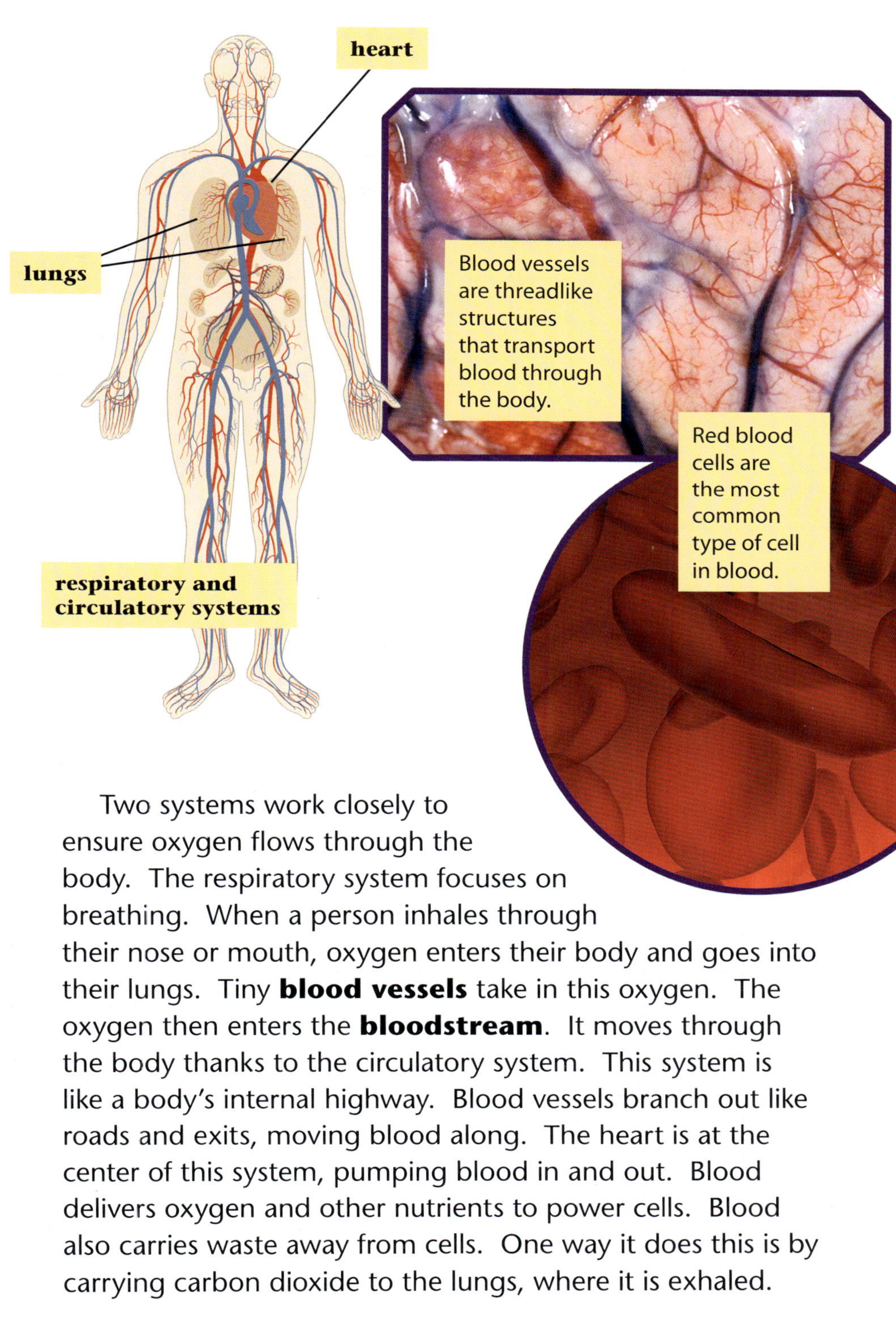

Blood vessels are threadlike structures that transport blood through the body.

Red blood cells are the most common type of cell in blood.

Two systems work closely to ensure oxygen flows through the body. The respiratory system focuses on breathing. When a person inhales through their nose or mouth, oxygen enters their body and goes into their lungs. Tiny **blood vessels** take in this oxygen. The oxygen then enters the **bloodstream**. It moves through the body thanks to the circulatory system. This system is like a body's internal highway. Blood vessels branch out like roads and exits, moving blood along. The heart is at the center of this system, pumping blood in and out. Blood delivers oxygen and other nutrients to power cells. Blood also carries waste away from cells. One way it does this is by carrying carbon dioxide to the lungs, where it is exhaled.

The nervous system is like the body's control center. It helps a person use their body and senses to react to the world around them. The brain, **spinal cord**, and nerve cells make up this system. They communicate back and forth to help a person react to things. Think about a person spraining their knee while playing soccer. Their nerve cells send an electrical signal through the spinal cord to the brain. The brain receives this signal and interprets it into a message, such as "Ouch! That hurt." Then, the brain sends a signal back to the nerve cells, telling them how to react: "Don't stand on that leg." This all happens quickly—in less than one second!

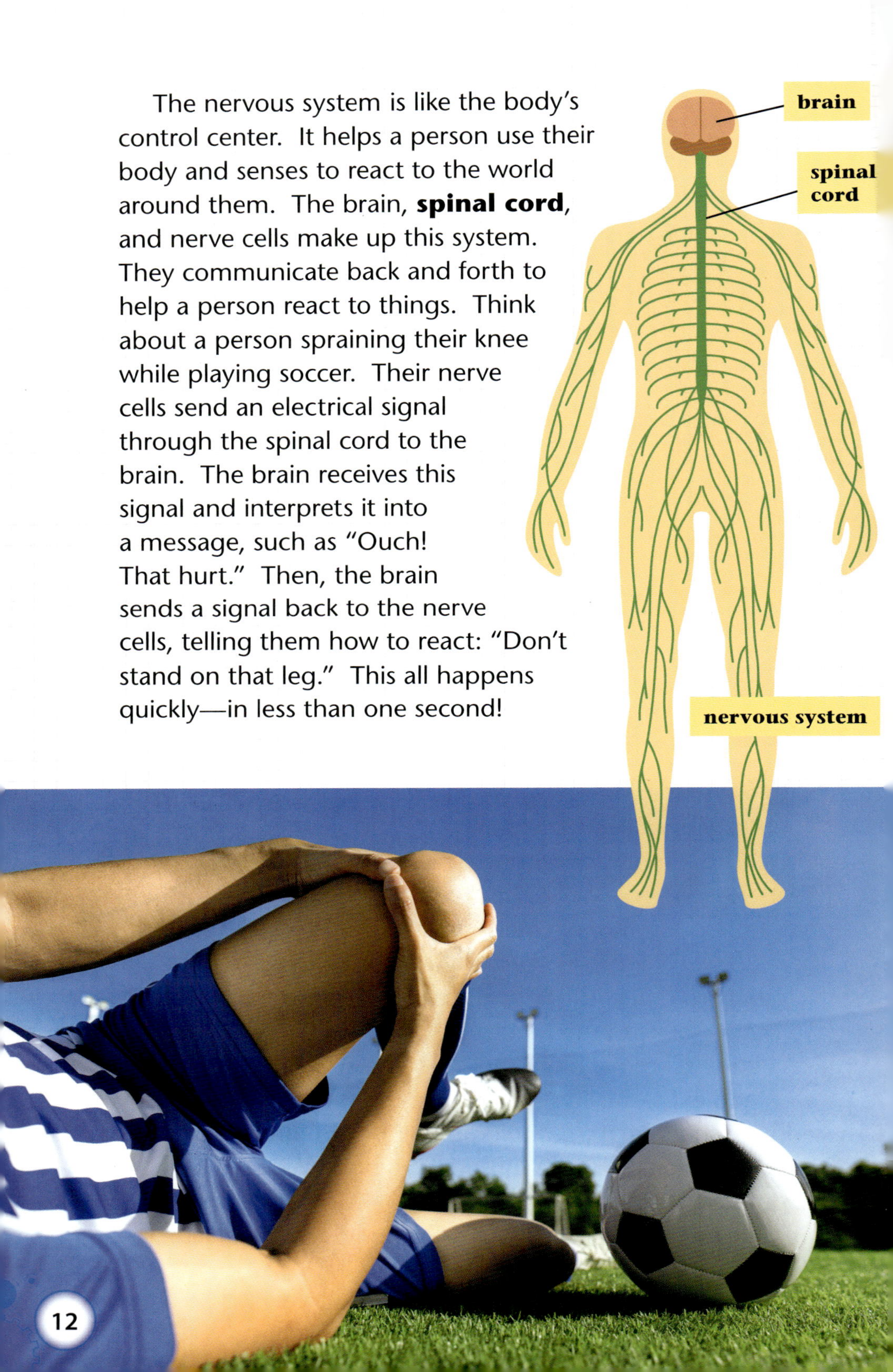

Transforming food into energy is a big job, and the digestive system takes care of this. It all starts in the mouth as food is chewed, mixed with saliva, and swallowed. The chewed food travels down the **esophagus** and into the stomach. Stomach acid breaks down the food even more, and within two hours, it goes into the small intestine. There, nutrients from the food are absorbed and delivered to the rest of the body. The remaining parts of the food are waste. Waste becomes **stool**, which exits the body through the large intestine.

The excretory system acts like a body's filter by removing substances that aren't needed. The kidneys are an important part of this system. They filter blood, removing any waste and extra water to make urine. Then, the bladder stores the urine until it leaves the body.

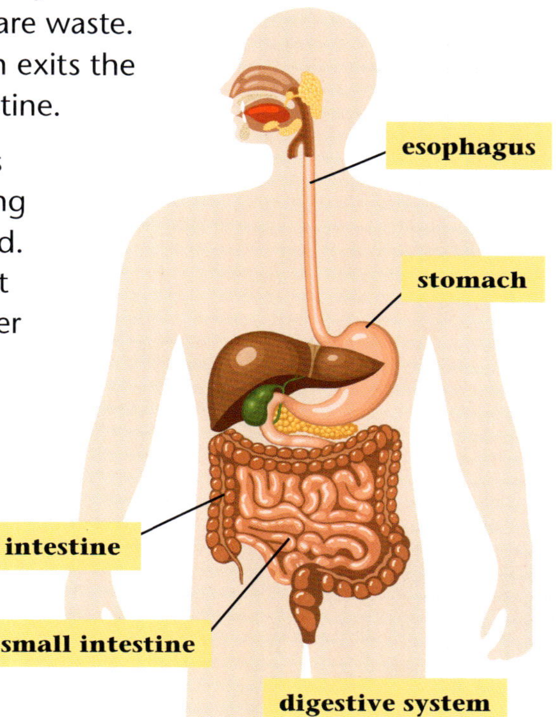

digestive system

FUN FACT

On average, every human body is made up of 60 percent water. So, it should be no surprise that staying hydrated is crucial for your body's systems. Drinking enough water puts less stress on your kidneys and liver. The right amount of water can also help fight and prevent headaches.

Chain Reactions: Responding to Invaders

Sometimes, the delicate balance among bodily systems can be broken. Pathogens, such as **viruses** and bacteria, can invade a person's body and disrupt one or more systems. These foreign organisms cause common illnesses, including the cold or flu. Pathogens gain access to the body through the nose, mouth, or skin. For example, a person might touch something with germs on it or breathe in infected air.

Viruses, such as these coronavirus cells, cause uncomfortable symptoms.

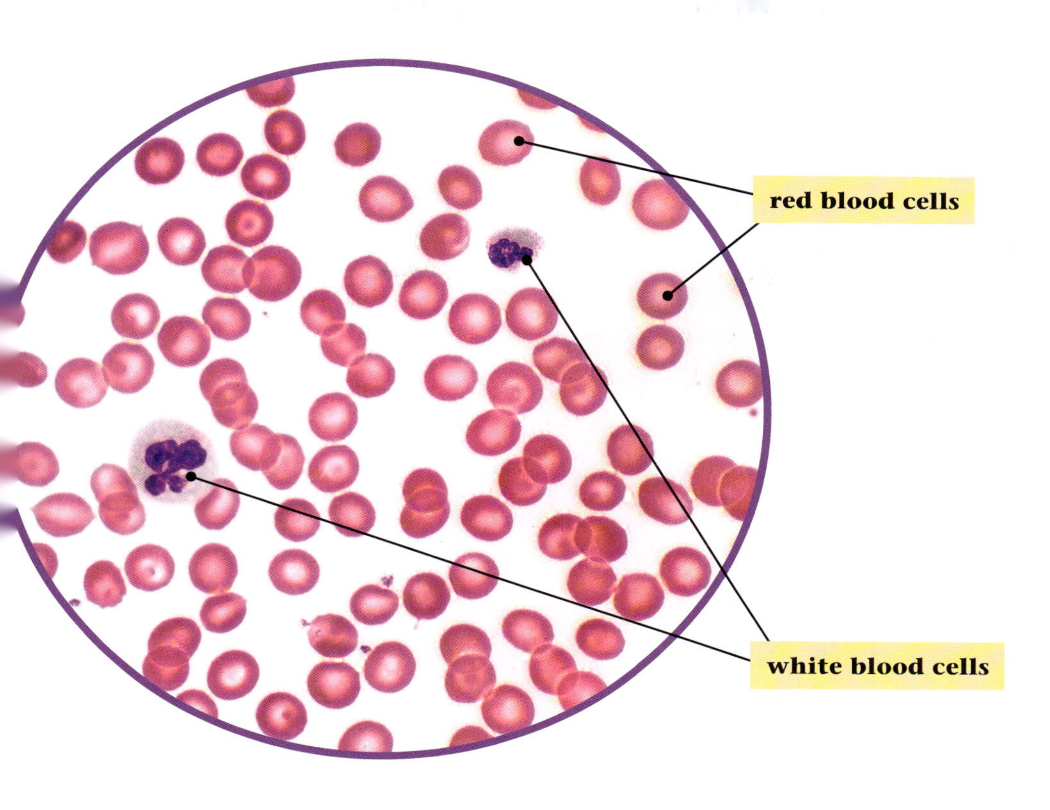

The body's response to an invader can cause a chain reaction. Consider the common cold, for example. It is caused by a virus. When this virus enters the body, the **immune system** responds. Blood vessels begin to swell. This swelling allows white blood cells to flow to the infected cells. White blood cells are an important part of the immune system. They protect a person against illnesses and diseases. A side effect of this swelling is **inflammation** in parts of the body. So, a person with a cold often experiences uncomfortable symptoms, such as a stuffy nose or a sore throat. Plus, the increase in blood flow causes fluids to collect in certain areas. This includes the nose and the lungs. A runny nose or a cough is the body's way of trying to release this extra fluid.

To reduce these symptoms, there are a few steps people can take. Getting plenty of rest, drinking lots of water, and eating healthy foods support the body's systems. These actions promote the strength and growth of cells.

Fighting and Recovery

As white blood cells fight off a virus, such as a cold, they begin training for future battles. They do this by collecting information about the virus. This information teaches the cells how to improve their counterattack the next time that virus invades. Having this information can prevent further illnesses. It can give a person natural **immunity** for a certain amount of time. However, there are hundreds of **strains** of viruses that cause the common cold. A person's body might learn to defend against one of them, but they could get sick from a different strain months later. That is why some people can get two or three colds per year.

Eventually, white blood cells wipe out the virus and begin to retreat. Swelling begins to reduce, and fluid buildup drains. The sick person's symptoms begin to fade, and they feel better. The entire recovery process can take anywhere from a few days to a few weeks.

common cold pathogens as seen under a microscope

Bacterial Infections

Just like viruses, bacteria can cause infection in a person's body. One example is food poisoning, which is caused by eating spoiled or contaminated food. Strep throat, an infection that results in a severe sore throat and high fever, is another example. In some cases, doctors may recommend taking antibiotics. These medicines work by killing bacteria. They also stop bacteria from multiplying. Antibiotics and a person's immune system work together to target bacteria. They help a person recover from a bacterial infection.

Antibiotics come in many forms, including pills, drops, lotions, and injections.

Doctors test for strep throat by swabbing a person's throat and tonsils for bacteria.

Rest and Rebuild: Healing Broken Bones

The average person experiences a broken bone at least once in their life. From falling off a bike to dropping something heavy on a toe, there are many ways broken bones occur. While the experience can be painful, it also showcases some of the most impressive work human bodies can do.

The moment a person breaks a bone, a complex response begins in their body. The skeletal and nervous systems work together to set off an alarm in their brain. The message says, "Something has gone wrong!" Certain cells release adrenaline, a special chemical. It makes the person's heart beat faster and increases their energy levels. The injured person likely feels pain, shock, or both. These changes prompt their body to start healing the injury.

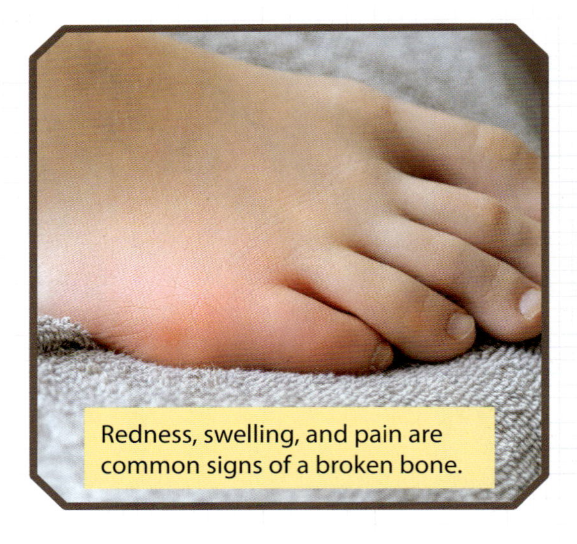

Redness, swelling, and pain are common signs of a broken bone.

After the initial break, the bone cells summon white and red blood cells to do damage control. Like the body's response to a virus, the rush of blood cells causes swellling. As a result, the injured area may appear red, and the person's skin may even feel hot to the touch. These symptoms, along with pain, alert the person to seek help.

TECHNOLOGY

A Glimpse Inside

X-rays allow doctors to take a closer look at a person's bones. These strong waves of energy move through the skin to take pictures of bones. In an X-ray, bones appear white and skin and soft tissues appear black. This helps doctors easily see where and how a bone has broken.

Medical technicians tell people how to place their injured limbs so that X-ray machines can take clear pictures.

Once there are enough blood cells near the broken bone, they unite around the injury to form a clot. The clot serves as a seal. It keeps the blood vessels inside the bone in place so they can heal. Then, tissue and **cartilage** form a shield called a *callus* around the break. The callus acts like glue as it holds the bone fragments together. Over time, the callus hardens. Special cells inside the callus chip away at the old bone, while other cells build new bone in its place.

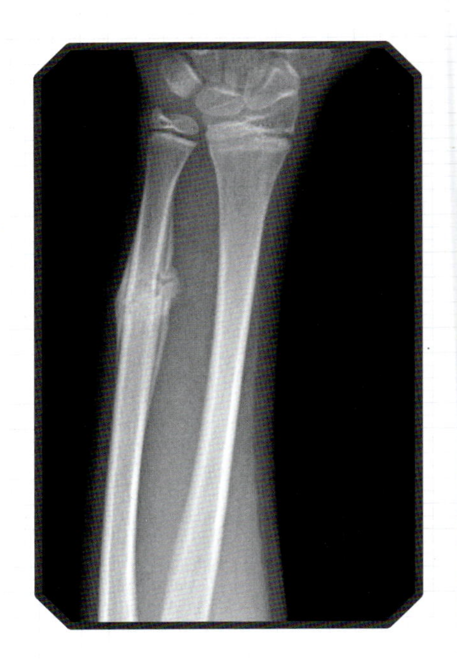

Rebuilding a broken bone is hard work! This microscopic construction crew can be slow-moving. A broken bone can take several weeks to a year to fully heal. It all depends on the severity of the injury.

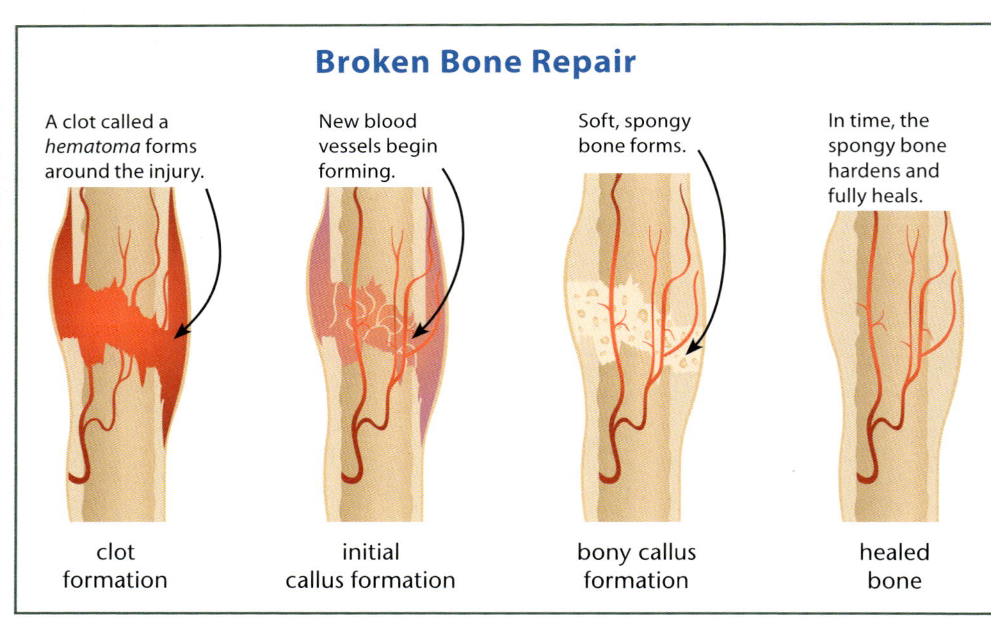

In the meantime, a person can support their recovery in a few ways. To start, staying hydrated and eating healthy foods can help their body recover. These actions are helpful because water and nutrients fuel cell growth and production. Bone cells especially thrive on calcium, a mineral that can be found in foods such as milk, cheese, and broccoli. Additionally, wearing a cast or a splint to keep from moving the injured area can help the bone heal correctly. If a person's splint can be removed, putting ice on the injured area can help, too. The coldness of the ice helps the blood vessels shrink to their normal size. This reduces the swelling and redness.

FUN FACT

Some people may wear a prosthesis. This is a human-made device that replaces a missing part of the body. Prosthetic legs and arms allow a person to gain movement back. Some scientists are working on developing bionic arms and legs. These use electronics to boost their abilities.

A boy uses a bionic hand to pick up blocks.

prosthetic arm

Everyday Occurrences

When a person gets sick or breaks a bone, a lot goes on in their body. The same is also true for everyday occurrences, such as scratching an itch, sleeping, and sneezing. These processes are more complex than you might think. They play important roles in keeping people's bodies healthy and safe.

Itchy and Scratchy

Have you ever felt itchy and just *needed* to scratch at your skin? An itch is one of the body's many ways of protecting itself. It alerts the body to potentially harmful agents trying to enter through the skin. Many outside **stimuli** can bring on an itch. Bug bites and rough fabrics are common culprits.

Feeling an itch and scratching it has several steps. First, an irritant touches the skin. Then, special itch-sensing nerve cells detect the sensation. Those cells send a message to other nerve cells in the body to say, "Something's itchy!" The message travels from the skin through the spinal cord, ultimately reaching the brain. The brain cells respond by ordering the muscles and bones in a person's arm and hand to scratch the itch. The scratching sensation tells the itch-sensing nerve cells that their message was received and addressed. The itchy feeling goes away, and the person experiences relief. Their body relaxes until another itch comes along.

mosquito

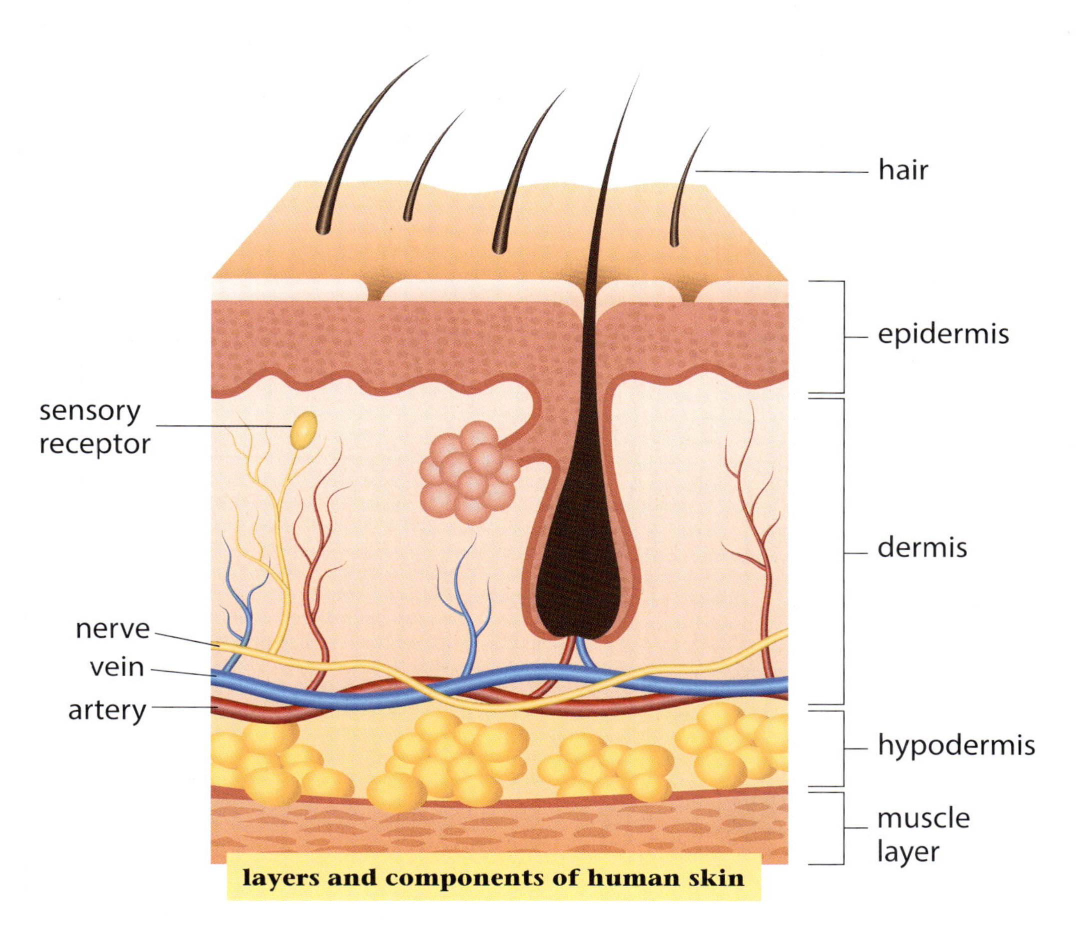

layers and components of human skin

ENGINEERING

Electronic Skin

Some smartwatches can monitor a person's heart rate and blood oxygen levels. But they can be big and bulky to wear. Engineers at MIT, a research university, came up with a solution. In 2022, they created an electronic "skin." The flexible film sticks to the skin like tape. Tiny sensors inside it can monitor pulse, sweat, and other body signals.

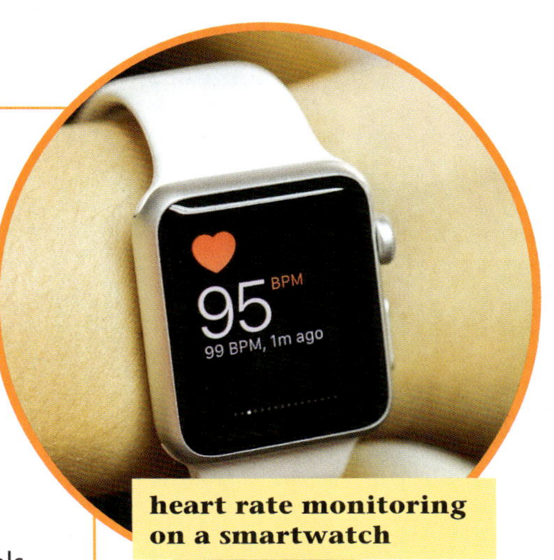

heart rate monitoring on a smartwatch

Sweet Dreams

Sleep is one of the most important processes for the human body. Although it might seem like nothing happens during sleep, your cells are hard at work. A good night's rest is like a reboot for cells.

When a person falls asleep, their heart rate and breathing slow down. Their muscles begin to relax until their whole body is fully relaxed. Periods of deep sleep allow cells to do some housekeeping. During this time, cells work on making proteins and removing waste. This quiet time allows them to repair, grow, and multiply.

Getting enough sleep is necessary for staying healthy. Lack of sleep makes cells and systems less efficient. Most adults need about six to nine hours of sleep to feel rested. Children and teens, whose bodies are still growing rapidly, may need to sleep even longer.

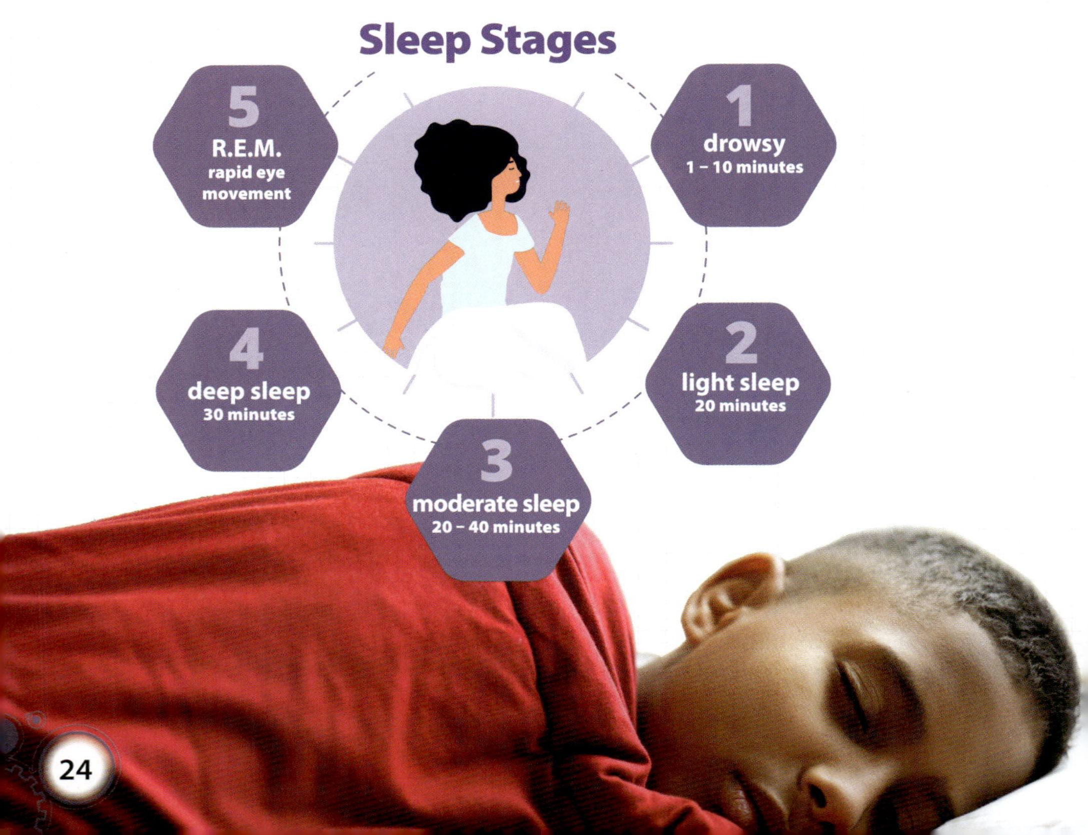

Achoo!

Sneezing is a common reflex among most humans. Like an itch protecting the skin, a sneeze protects the lungs. When an irritant enters a person's nose, their nerve cells send off an alarm message. The message tells their brain, "An intruder is trying to get in!" Their brain broadcasts this message to their lungs and mouth. As a result, the person takes a deep breath, and their lungs inflate. Then, their tongue presses against the roof of their mouth. Air is forced out through their nose in a powerful rush. This sudden blast of air blows the intruder away. Invasion averted!

SCIENCE

Eyes Shut Tight

Have you ever noticed that you close your eyes when you sneeze? Some scientists think it is another way the body tries to protect itself. Closing your eyes stops an irritant in the nose from going up into your eyes.

Beyond the Basics: Keep Discovering

Human bodies are complex machines. Microscopic cells come together in tissues, organs, and systems. These intricate systems keep us alive, healthy, and growing. When all systems work correctly, they help our bodies perform countless tasks every minute of every day.

Every process that occurs in the human body involves elaborate work. In times of illness, the body launches a powerful defense against invaders. The immune system responds by fighting back. It stores data to make its defenses stronger in the future. In times of injury, the skeletal and nervous systems work together. Cells take on the tough job of rebuilding bone. Even in everyday events, the body shows off its amazing abilities. Scratching an itch prevents potential harm. Sleeping allows cells to repair themselves. Sneezing protects the lungs from intruders. Each of these processes supports and protects the body.

The more we learn about the functions and abilities of our bodies, the better we can care for ourselves. Basic knowledge about cells and systems is just the start. As you keep growing, you'll find that there is an entire world of information to discover!

ARTS

Permanent Body Art

Some people put artwork on their bodies in the form of tattoos. Tattoo artists use a sharp needle to inject ink under the top layer of a person's skin. The ink stays there, creating a permanent design. As you might imagine, tattoos can be painful. When the needle punctures the skin, the nervous system sends pain signals between the skin and brain.